LOUISA COUNTY [VIRGINIA] ROAD ORDERS

1742-1748

Virginia Genealogical Society
Richmond, Virginia

Published With Permission from the

Virginia Transportation Research Council
(A Cooperative Organization Sponsored Jointly by the Virginia
Department of Transportation and
the University of Virginia)

HERITAGE BOOKS
2008

HERITAGE BOOKS

AN IMPRINT OF HERITAGE BOOKS, INC.

Books, CDs, and more—Worldwide

For our listing of thousands of titles see our website
at
www.HeritageBooks.com

Published 2008 by
HERITAGE BOOKS, INC.
Publishing Division
100 Railroad Avenue #104
Westminster, Maryland 21157

International Standard Book Number: 978-0-7884-3661-1

LOUISA COUNTY ROAD ORDERS 1742-1748

by

Nathaniel Mason Pawlett
Faculty Research Historian

Virginia Highway & Transportation Research Council
(A Cooperative Organisation Sponsored Jointly by the Virginia Department
of Highways & Transportation and the University of Virginia)
Charlottesville, Virginia
April 1975
Revised January 1979
Revised August 2003
VHTRC 75-R43

Library of Congress Catalogue Card
No: 79-620003

PREFACE

The Virginia Highway and Transportation Research Council is a cooperative organisation sponsored jointly by the Virginia Department of Highways and Transportation and the University of Virginia and is located on the Grounds of the University at Charlottesville. The Council engages in a comprehensive program of research in the field of transportation. As a part of its program the Council, in December 1972, began research on the history of road and bridge building technology in Virginia. The initial effort was concerned with truss bridges; a complementary effort concentrating on roads got underway in October 1973.

The evolution of the road system of Virginia is in many ways inseparable from the social, political and technological developments that form the history of the Commonwealth. Despite this, there are few extant serious works on the history of roads in Virginia. Those which have been produced focus on internal improvements and turnpike development before the War Between the States. Little has been done on the period from Reconstruction through the creation of the system of state highways in the earlier part of this century.

Accordingly, it was decided to investigate the development of the roads of Albemarle County during the period 1725-1925 as a pilot project, and to use this experience to produce a "History of Albemarle County Roads" and a procedural handbook for the writing of Road Histories. During the early stages of this project it was necessary to *examine* and extract all the Road Orders for the Counties from which Albemarle was formed as well as the Orders for Albemarle when it still contained the Counties of Amherst, Buckingham, Fluvanna, Nelson, and a part of Appomattox. The Road Orders concerning Albemarle will ultimately be published with the Road History but the broad applicability of those for Goochland, Louisa and early Albemarle, and the opinions of various authorities throughout the state who have examined them, indicate that they should have separate publication in order to make them generally available to individual scholars through libraries and educational institutions. Therefore, this publication has been prepared.

LOUISA COUNTY ROAD ORDERS 1742-1748

by

Nathaniel Mason Pawlett
Faculty Research Historian

INTRODUCTION

The roads are under the government of the county courts, subject to be controuled by the general court. They order new roads to be opened whenever they think them necessary. The inhabitants of the county are by them laid off into precincts, to each of which they allot a convenient portion of the public roads to be kept in repair. Such bridges as may be built without the assistance of artificers, they are to build. If the stream be such as to require a bridge of regular workmanship, the court employs workmen to build it, at the expence of the whole county. If it be too great for the county, application is made to the general assembly, who authorize individuals to build it, and to take a fixed toll from all passengers, or give sanction to such other proposition as to them appears reasonable.

Thomas Jefferson, *Notes on the State of Virginia*, 1781.

The establishment and maintenance of public roads was an important function of the County Court during the colonial period in Virginia. Each road was opened and maintained by an Overseer or Surveyor of the Roads charged with this responsibility and appointed by the Gentlemen Justices. He was usually assigned all the "Labouring Male Titheables" living on or near the road for this purpose.

Major projects, such as bridges over rivers, demanding considerable expenditures were executed by Commissioners appointed by the Court to select the site and contract with workmen for the construction. Where bridges connected two counties, a commissioner was appointed by each and they cooperated in executing the work.

The Road Orders contained in the first Louisa County Court Order Book covering the period 1742-1748 are the only extant evidence concerning the early roads of the County since there are no surviving Order Books for the period 1748-1766 and the early Hanover records have been destroyed. Approximately the upper third of the present Albemarle County was in Louisa until 1761. Before this the Louisa line extended from about where the Fluvanna-Louisa line intersects the eastern boundary of Albemarle in the northwesterly direction, across the point where Ivy Creek enters the Rivanna River, to the crest of the Blue Ridge just south of Brown's Cove. Before 1742 this area had been the westernmost part of Hanover County.

In the following Road Orders those relating to the present configuration of Albemarle County are marked with an asterisk. Insofar as possible, all the Orders were extracted <u>verbatim</u> and the capitalization, spelling and punctuation have been reproduced without any attempt at correction or consistency.

THE DEVELOPMENT OF LOUISA COUNTY

Note: As originally published in paper format, this volume included maps showing the evolution of the county. Maps are not included in the revised/electronic version due to legibility and file size considerations. Instead, a verbal description is provided.

By the 1720s, the area that is now Louisa County was part of the western reaches of Hanover County (created in 1721 from New Kent County). In 1742, the western section of Hanover was cut off as Louisa County.

As originally created, the western section of Louisa County included the northern third of modern-day Albemarle County. In 1762, this region was cut from Louisa County and added to Albemarle County. With this change, Louisa County's boundaries reached their present locations.

Louisa County Order Book 1742-1748

At a Court for Louisa County…

13 Dec. 1742 Old Style, Page 5

Ordered That all Surveyors of the Roads in the County appointed by Hanover Court be continued as such until it shall be otherwise ordered by this court and that the Sherif give public notice thereof.

Benjamin Dumas is appointed Overseer of the Road in this County where of Joseph Bickley was appointed Overseer by Hanover Court

* 10 January 1742 Old Style, Page 7

At the motion of Samuel Dalton It is ordered that he be overseer of the Road from the Mountains to Hudson's bridge including the s^d. bridge and no further.

Ordered that that part of the Road from Hudson's bridge to Camp-creek bridge be added to the Road whereof Charles Moreman is Overseer.

14 February 1742 Old Style, Page 10

John Rhodes and David Cosby at the motion of Charles Barret Gent: are appointed to view the way from Stephen Pettus's plantation to John Estes's Race Ground and report to the next court where the most convenient road may be made.

* 14 March 1742 Old Style, Page 17

On the petition of Henry Bunch, John Snow, Samuel Wells, Joseph Keaton, James Keaton, John Enes, Alexander MacKoy, Martin Hacket, John Hacket, William Davis, Thomas Vernon James Frances, Beasell Maxwell, James Carney, Andrew Rey, John Cook, John Thompson and William Oens It is ordered that a Road be laid off and Cleared from Coursey's Road up to Rocky Creek and Henry Bunch is appointed Overseer of that part above Buck-mountain Creek and John Snow of that part below the same And all the petitioners and their Labouring Male tithables that live above Buck mountain Creek are allotted to Henry Bunch and those below to John Snow to assist in clearing the s^d. Road and keeping the same in repair.

14 March 1742 O.S., Page 18

John Rhodes and David Cosby being appointed last Court to view the way from Stephen Pettus's to John Estes's Race Ground and to report where the most convenient Road might be made, and now reporting that they find the s^d. Road must be carryed along William Harris's Cartway over the Horse-pen swamp, and then to follow some blazed trees to the left hand till it comes to the Mill path, and then to follow the same across the River just above the Old Dam and from thence to keep in or near a path that leads up to the s^d. Race Ground: It is therefore Ord. that a Road be laid off and cleared accordingly, And James Nuckols is appointed

Overseer thereof, And the labouring Male Tithables belonging to Charles Barret Gent, John Estes, Nicholas Gentry, John Cosby Quart.^{r}, William Harris, Richard Yancey, James Nuckols, & John Henson are allotted the s^d. Overseer to assist in clearing the s^d. Road and keeping the same in repair.

* 14 March 1742 O.S., Page 18

On the motion of Ambrose Joshua Smith Gent. It is ordered that William Coursey's Company on the Road be divided as follows, be the s^d. Wm. Coursey to keep all the hands on the North side of the North fork of James River & to keep that part of the s^d. Road in repair which leads from Goochland County line to the branch next below the s^d. Smith's plantation, And that the hands between the North side the branch called Turkey-run & Orange County line to keep the Road in repair from the aforesd. branch next below the s^d. Smith's plantation to the extent of the County Road of which William Ogilvie is appointed Overseer.

* 14 March 1742 O.S., Page 19

Gideon Carr is appointed Overseer of the Road whereof William Maybe was formerly Appointed Overseer by Hanover Court.

14 March 1742 O.S., Page 19

On the motion of Lancelot Armstrong Overseer of a Road It is ordered that the labouring Male Tithables belonging to William Gooch, & to Doct^r. Dixon where David Woodroof is Overseer, also to Mr. Wm. Gray where Benjamin Spencer is Overseer, do assist in keeping the s^d. Road in repair.

14 March 1742 O.S., Page 19

William Allen is appointed Overseer of the Road from the County line to that part of the Road whereof Thomas Christmas was Overseer, also to that part of the Road whereof Anthony Waddy was Overseer, And the labouring Male tithables at the Quarter where Christopher Berryman is Overseer, and at Doct^r. Dixon's quarters where Benjamin Johnson & James Whitlock are Overseers, and also at his Fork Quart^r. are allotted the s^d. Overseer to assist in keeping the s^d. Roads in repair.

11 April 1743 O.S., Page 26.

John Kendrick is appointed Overseer of the Road from the Elk Creek to contrary including the bridge.

11 April 1743 O.S., Page 26

William Taite is appointed Overseer of the Road from Contrary to Christophers Run including the bridge.

11 April 1743 O.S., Page 26

     Samuel Thompson is appointed Overseer of the Road from Christopher's Run to Foxbranch on Hickory Creek and the labouring Male tithables at Mr. Temple's and Maj^r. Morris's quarters and the s^d. Thomsons two hands which used to serve on Champnes Terry's road are added to his Company to assist in keeping the s^d. road in repair.

11 April 1743 O.S., Page 27

     Champnes Terry is appointed Overseer of the road from the bridge over the North Anna into the Road at Ellis Hughes's and the labouring Male Tithables belonging to John Chiles, William Biggars, & Wm. Thomason are added to his Company to assist in keeping the s^d. road in repair.

11 April 1743 O.S., Page 27

     James Robinson is appointed Overseer of the Road from the mouth of Starke's branch to Aylett's branch and the labouring Male tithables belonging to Thomas Starke are added to his Company to assist in keeping the s^d. Road in repair.

11 April 1743 O.S. Page 27

     On the petition of Richard Bullock, Harper Ratcliff, Edward Lane, Robert Garland, Benjamin Brown, James Scolson, and Richard Yancey, praying for a Road from little River road against the head of Great Rocky Creek & thence crossing the little River at the most convenient place near a Foard called the Rocky Foard thence to Mr. Barret's Church path on the North side of Rocky branch, David Cosby and John Rhodes are appointed to view the place and layoff a Road the most convenient way without prejudice to any person and report their proceedings to the next Court.

11 April 1743 O.S., Page 27

     John Gentry is appointed Overseer of the Road in the stead of James Nuckols, And French Haggard & John Saxon are added to his company to assist in repairing the s^d. road in the stead of John Estes and Nicholas Gentry who are discharged from that Road.

13 June 1743 O.S., Page 34

     On the Petition of Charles Mooreman overseer of a Road Forest Green is appointed in his stead.

13 June 1743 O.S., Page 34

     On the Petition of Richard Wright Overseer of A Road Thomas Lipscomp is appointed in his stead.

12 September 1743 O.S. Page 73

     Upon the motion of Abraham Venable Gent: it is ordered that Owens Creek bridge be added to Mr. Thompson's precinct.

12 Sept. 1743 O.S., Page 73

Ordered that William Brown be Surveyor of the road from Owen's creek bridge to the County line and all the inhabitants between Owens creek & the County line on the South Side of the River be added to Mr. Thompsons Gang towards clearing the said road.

10 Oct. 1743 Page 80

On the petition of Anne Johnson Love Statham James Watson & Peter Curry for a road to be cleared into a road called Christmas's road in this County its Ordered that the said road be cleared according to the prayer of the petitioners and Love Statham is appointed Overseer and all the Labouring male tithables belonging to the said Love Statham Anne Johnson & Richard Wilburn are allotted to assist in clearing the said road and keeping the same in repair.

20 Oct. 1743 O.S., Page 80

John Estis is appointed Surveyor of the highways in the room of Richard Estis deceased.

* 10 Oct. 1743 O.S., Page 80

On the petition of Robert Thompson David Epperson Martin Hackett George Pearse Richard Bennet Richard Madors its ordered that a road be laid of and cleared from the road in Orange that extends to the dividing line between this County and Orange on Linches river to the upper north fork of Buck mountain creek along the track that leads to Robert Thomson & Martin Hacket is appointed Overseer and all the pet^{rs}: & their Labouring male tithables are allotted to assist in clearing the said road and keeping the same in repair.

19 Dec. 1743 O.S., Page 84

Thomas Hardy is appointed Overseer of the Church Road Down to the County Line.

19 Dec. 1743 O.S., Page 85

Joseph Clarke is appointed Overseer of the Road from Fork Creek to Deep Creek and all the Labouring Male tithables between the said Creeks are allotted to assist in clearing the said Road and keeping the same in repair.

19 Dec. 1743 O.S., Page 85

Upon the petition of Henry Mills and others, Setting forth that they want a road from Elk Creek and Little River in this County Ordered that Griffey Dickerson John Roads and Robert Estis do view the same and report their proceedings herein to the next Court.

* 19 Dec. 1743 O.S., Page 85

Upon the motion of John Carr Gent. to have a road to go out of (blank in O.B.) road to Mountaine road through Turkey Run Saggs and into the road the West side the mountains it is ordered that

Benjamin Hensley Joseph Martin and John Dickerson meet at Some convenient before the next Court, and view the same, and make report of their proceedings herein to the said next Court.

9 Jan 1743 O.S., Page 89
Ordered that the Lands from Gilbert Gibsons Mill creek up to Orange County line be added to Benjamin Johnson's precinct.

9 Jan 1743 O.S., Page 89
Ordered that the Lands from Christophers Run to Contrary Run and between the north River and Ridge Road be added to James Estis precinct.

* 9 Jan. 1743 O.S,. Page 89
John McCawley is appointed overseer of the road in the room of Gideon Carr.

13 Feb. 1743 O.S., Page 93
Ordered that Mr. Terrils Richard Thormond Cleverius Dukes Mr. Anderson & Thomas Jones hands be added to Thomas Hardy's Gang.

13 Feb. 1743 O.S., Page 94
On the petition of Joseph Martin & Benjamin Hensley for a road, Ordered to be referred till John Carr Gen: be heard.

13 Feb. 1743 O.S., Page 95
On the petition of Henry Mills & others for a road it is ejected.

13 Feb. 1743 O.S., Page 97
On the petition of Nathaniel Dickenson Adam Jones Jn°: Harris & others for a Road that leads from the Mine road over the Little River to Locust Creek road it is Ordered that Richmond Terrill be Overseer of the said Road & that all his Labouring male tithables belonging to the s^d. Richmond, Richard Thormons John Moss Adam Jones, James Sims are allotted to assist in clearing the said road and keeping the same in good repair and the aforementioned persons are to keep the County road from the Line to Duke upper Line.

12 March 1743 O.S., Page 98
On the Petition of Charles Allen for a License to keep an Ordinary at his house on the Three notch road in this County and he offering Thomas Paulett for his Security whom the Court doth approve order him Granted him for the said License to continue and be of force for one year only from the date of this order,

12 March 1743 O.S., Page 98

On the Motion of George Webb Gen: to have a road from Roundabout Creek to Fosters Creek in this County', it is ordered that John Smithson Henry Tate and John Davis do meet at some convenient time before the next Court and view the Said road and see wether its convenient or not and report their proceedings herein to the said next Court.

12 March 1743 O.S., Page 98

On the petition of Mosias Jones William Thompson James Merideth & Robert Thompson Setting Forth that the road from the River to the top of the Hills below the Little creek is very hilly & that a much better & nearer way may be had & &. and praying to be relieved it is Ordered that David Mills & John Cook do meet at some convenient time before the next Court and view the same and Report wether it is convenient or not to the next Court.

12 March 1743 O.S., Page 98

On the Petition of David Meriwether Gent. to have a road from Goodalls Lower ford to Henson's Creek church, it is Ordered that Mr. Glen Mr. Jackson & Mr. Thompson do meet at some convenient time before the next Court and view the same & make report of their Proceedings herein to the said next Court.

12 March 1743 O.S., Page 99

William Ogilvie on his Petition is discharged from being any Longer Surveyor of the highway and Richard Durret is appointed in his Room

* 12 March 1743 O. S., Page 99

On the Motion of John Carr Gen: to have a road to begin at the Chestnut Ridge it is ordered that John Meriwether & Doct^r. Walker do meet at some convenient time before the next Court and Lay off the said road accordingly.

12 March 1743 O.S., Page 99

On the Motion of John Poindexter Gen: to have a way viewed from Estis race Ground into the north River road called by the name of Overton's Quarter, it is Ordered that John Roads Charles Bickley & William Saxon & John Estis do view the same & make report of the most convenient way to the next Court.

12 March 1743 O.S., Page 99

Ordered that Richard Palmer be Overseer of the Indian Creek Road in the room of William Sled.

9 April 1744 O.S., Page 102

On the Petition of Mosias Jones Willm: Thompson James Merideth Robert Thompson for a Road to be laid off etc. it was Ordered that David Mills & John Cooke should view the same & make report wether it was convenient, which report is in the following words. "Pursuant to an order of, Louisa County Court bearing date the 12 Day of March 1743 We have viewed & marked away which is better & nearer than the way the road is P David Mills John Cook" Therefore it is Ordered that the Said road be decryed & taken as a Publick road for the future

9 April 1744 O.S., Page 102

The Return of a view for a road at the request of John Poindexter is rejected

9 April 1744 O.S., Page 102

On the Petition of John England John Carr Jno: Poindexter & divers others for a road on the North River Road by the Plantation that was formerly James Overton's Quarter along the ridge between the South & North forks of Elk Creek & crossing the branches of Contrary to the ridge, between the branches of Christophers' Run, Ducking hole, Gold Mine, and the branches of East & Bever Creeks into the Mountain Road, it is Granted and Griffith Dickenson is appointed Overseer of the said Road.

9 April 1744 O.S., Page 103

On the Petition of David Meriwether Gen: for a road, it was Ordered last Court that Mr. Glen & Mr. Jackson should view the same & make report to the next Court. And this day the Report was re- turned which is in the following words "Pursuant to the within order we have viewed the way and find there may be a rode had from Cattale bridge in Blalocks rode to Church and that it is convenient Jer: Glen Thomas Jackson" Therefore it is considered & accordingly ordered that there be a road from the place aforesaid according to the s^d. report.

9 April 1744 O.S., Page 104

On the Motion of George Webb Gent for a road to be laid off from Roundabout Creek to Foster's Creek, it was ordered at a Court held for this County on the 12 day of March last, that John Smithson Henry Tate & John Davis should view the same & make report to the next Court, and this day the s^d. report was returned being in these words "In obedience to the within order we have viewed the Road & do find that it is convenient for Mr. Webb the new way being by Computation Six Miles out of his way, ye. road that Mr. Webb motions for is seven Miles & has Three Creeks, the Gang could not keep bridges over without a County charge. John Smithson Henry Tate & John Davis Therefore it is considered by the Court & accordingly ordered that there be a road cleared according to the s^d. report.

14 May 1744 O.S., Page 106

Ordered that William Allen Surveyor of the Road from the County Line to that part of the road whereof Thomas Christmas was Overseer and to the head of Long Creek be summoned to appear at the next Court to answer the Presentment of the Grand Jury this day made against him for not keeping his roads in Good repair according to Law

11 June 1744 O.S., Page 108

On the Petition of Richard Bullock and others for a road from Little River road against the head of Great rocky creek and thence crossing the little River at the most convenient Place near a Foard called the rockey Foard, thence to Mr. Barrets Church path on the north side of rocky branch, John Rhodes and Samuel Mcgehee are appointed to view the Place and lay of a road the most convenient way without Prejudice to any Person and report their proceedings to the next Court.

* 11 June 1744 O.S., Page 108

The order for John Meriwether & Thomas Walker Gen: to lay off a road from the chesnut ridge etc. not being performed the same is continued till next Court.

11 June 1744 O.S., Page 108

Charles Bickley is appointed Surveyor of the highways from Elk creek to contrary including the bridge in this County in the room of John Hendrick

11 June 1744 O.S., Page 108

Samuel Ragland is appointed Overseer of a road (called Elk creek Road) in the room of Samuel Mcgehee.

11 June 1744 O.S., Page 108

Wm. Allen case continued

11 June 1744 O.S., Page 108

John Rhodes is appointed Overseer of the road from Elk Creek to the Church in the room of Henry Bibb

11 June 1744 O.S., Page 108

Ordered that the Several Surveyors of the highways in this County (except those discharged today) be continued as they were last year and where two or more crossroads or highways meet they forthwith cause to be erected in the most convenient place, where such ways join a stone or Post with inscriptions thereon in large letters directing to the most noted Place, to which each of the Said joining roads Leads

9 July 1744, Page 112

On the petition of Richard Bullock Sr. for a road continued till the next Court

9 July 1744, Page 112

On the Petition of Joseph Bickley Gen: setting forth that the Bridle road to the Church at Hinsons creek in St. Martin's Parish Runs very near his fence on one side of his Plantation and other roads on the other side so that it is very prejudicial to him and praying to have the said Path turned farther off from his plantation it is Ordered that John England William Davenport Samuel Mcgehee do meet at some convenient time before the next Court and view the same & make report of their Proceeding herein to the sd. next Court.

9 July 1744 O.S., Page 113

On the motion of William Allen, it is Ordered that Charles Barret & Joseph Fox Gen: do meet at Some convenient time before the next Court & view the road from Locus creek road to Christmas's road & make report of their Proceedings herein to the s^d. next Court

9 July 1744 O.S., Page 113

On the motion of William Allen, it is ordered that Richard Walker Michael Smith and Robert Harris & the Labouring male Tithables belonging to each of them are allotted to the said Allen's Gang.

9 July 1744 O.S., Page 113 d t

Grand Jury vs. William Allen, this day the s^d.Def^t.appeared & being fully heard it is ordered that the suit be dismist

13 August 1744 O.S., Page 115

On the petition of Joseph Bickley Gen: setting forth that the Bridle road to the church at Henson's creek in St. Martin's Parish runs very near his house on one side of his Plantation & other roads on the other side; so that it is very Prejudicial to him & Praying to have the said Path turned, it was Ordered at the last Court that John England William Davenport & Samuel Mcgehee should view the same and this day the viewers aforesd. have made their report, which report is in the following words "Aug. 10th: 1744 "Persuant to the within Order we the subscribers have viewed the within mentioned rode and do find that it must begin at the burnt School house crossing the head of Cabbin branch thence down the s^d. branch to Arche:^{s} Yancey Path where it crosses the branch below Bickleys Plantation thence into the former road: Jno: England, Wm. Davenport, Sam: MacKgehee" Therefore it is considered and accordingly Ordered by the Court that the Path be turned according to the viewers report.

13 August 1744 O.S., Page 116

On a motion of William Allen for a road the order made at the last Court not being Performed it is Ordered that Joseph Fox &

Charles Barret Gen: do meet at some convenient time before the next Court and Perform the same & make report of their Proceedings to the s^d. next Court .

* 8 October 1744 O.S., Page 119
On the Petition of James Merideth, Samuel Wells & John Garrison for a road to be cleared from Jones's road along the ridge to the green Mountain and that the Petrs: be appointed a Gang to clear the same; therefore, it is ordered that the Pet^rs. clear the road according to the Prayer in the Petition. and Samuel Wells is appointed Overseer of the said road

8 Oct. 1744 O.S., Page 119
Ordered that Charles Bickley Surveyor of the road from Elk creek to contrary have for his Gang Henry Bibb, John Poindexter's quarter above the said creek Mr. Winstons two Quarters above ditto & the Governor's Quarter where David Cosby now lives

8 Oct. 1744 O.S., Page 119
Ordered that the Road out of the north River road to Mills Path there be appointed under Griffith Dickeson Overseer, the following hands Viz^t. John Carr, John Poindexter, Joseph Terrell, David Cosby, William Saxon, James Overton, Armistead Quarter Under Robert Ellis Overseer of the same road Smiths Quarter, John M^cdowel, Thomas Gresham, Moses Estis, Robert Estis, James Johnson, William Hendrick, John Shelton's quarter, Edward Stringer, James Hower's quarter, William White Maj.^r Morrs's quarter & John Say; and it is Ordered that they be exempted from all other roads.

8 Oct. 1744 O.S., Page 120
On the Motion of William Allen for a road the order made at a Court held for this County.on the ix^th day of July last not being Performed it is Ordered that Charles Barret & Joseph Fox Gent: do meet at some convenient time before the next Court and Perform the same

* 12 November 1744 O.S., Page 122
Ordered that a road be cleared from the old Mountain road near the old chaple into Capt: Clark's road; thence up to Capt: Thomas Meriwether's Smiths' Shop; thence a long the same to Goochland County line; and all the Labouring male tithables belonging to Collo: Robert Lewis Capt: Clark; Micajah Clark; Anthony Tate, John Harlow on the Bever damm fork and Mr. Meriwethers' are allotted as a gang to clear the same & to Keep it in good repair & Thomas Meriwether Gen: is appointed Overseer of the same

* 12 Nov. 1744 O.S. Page 123
Ordered that John Snow be summoned to appear at the next Court to be held for this County on the xxiid day of January next

to answer the Presentment of the Grand jury this day made ag^t: him for not keeping his road in good repair according to Law.

* 22 January 1744 O.S., Page 128
John Snow upon his Motion is discharged from being any Longer Surveyor of the highways from Henry Bunches road at Buck Mountain Creek in this County to Coursey's Road and Andrew Rae is appointed surveyor of the said highway in the room of the said John Snow.

* 22 January 1744 O.S., Page 131
John Snow this day appeared to answer the Presentment of the grand jury made against him for not keeping his road in good repair, and what he could Say in excuse for himself being fully heard; it is considered by the Court the s^d: Presentment be dismist.

26 February 1744 O.S., Page 134
Upon the Petition of Nicholas Gentry & Samuel Gentry Setting forth that the road below dirty Swamp as it now Runs is very prejudicial to them, and they having cleared a new Road; turning out between their Plantations and thence into the road about three quarters of a Mile below; and Praying that the said road May be turned the new way &^{C}. it is Ordered that the road now cleared by the Petitioners be deemed & taken as a public road.

26 Feb. 1744 O.S., Page 134
Upon the Petition of William Hudson Setting forth that he is in want of a road through Capt. Holland's Land into Gibson's road he having no outlett to carry his Tobacco to the Public Warehouse & praying that a road may be cleared as aforesd: it is Ordered that a road be laid off & cleared according to the Prayer of the Petitioner

* 26 Feb. 1744 O.S., Page 135
On the Motion of John Carr Gent: to have a road to begin at the Chesnut ridge; at a Court held for this County on the xith day of June last; it was Ordered that Thomas Walker & John Meriwether Gent. should meet and lay off the said road and make report to next Court; and now this day the viewer made their report as follows & to wit "In obedience to the within order Wee have viewed & laid off the road therein Mentioned John Meriwether Thomas Walker" Therefore it is Ordered that the said road be cleared through the Turkey Sags; and that the hands belonging to Madame Meriwether Nicholas Meriwether Thomas Walker John Meriwether Robert Lewis Thomas Meriwether & Christopher Clark Gent: are allotted to clear the same and Andrew Tate is appointed Overseer Likewise the hands that William Coursey is Overseer of Richard Durret & John Mccollay to clear from his road; to meet the other gang on the Top of the mountains & Richard Davis is appointed Overseer of the Same.

* 23 April 1745 O.S., Page 145

Upon the Motion of Richard Hammack it is Ordered that John McCaulay's, Richard Durret's, William Coursey's and Thomas Ballard's Gang do make a bridge over Pritty's Creek below Ambrose Joshua Smiths Mill

* 23 April 1745 O.S., Page 146

Upon the Motion of Richard Hammack Benjamin Hensley John Henesley George Davis Gideon Carr Timothy Moony Stephen English John Moony John Hammack Charles Smith; John Maccaulay Joseph Martin Samuel Hensley William Maib Richard Davis; John Wilmer James Hoderurds John Wood & John Dowell for a road to be cleared from the s^d: Richard Hammack's Plantation to Ambrose Joshua Smith's Gent. It is Ordered that the Petitioners clear the same.

28 May 1745 O.S. Page 151

Charles Smith James Yancey and Archelau Yancey at the Motion of Thomas Prestwood are appointed to view the way from the road that crosses at Dickerson's Mill to the new church and report to the next Court where the most convenient road may be had

28 May 1745 O.S., Page 152

Ordered that the Surveyors of the roads from fork creek to Owens Creek, And overly dirty Swamp by Richard Haggard's, and Samuel Dalton to be summoned to appear at the next court to answer the Presentment of the grand Jury this day made against them for not keeping their Roads in repair according to Law.

28 May 1745 O.S., Page 152

On the Petition of John Estis, it is ordered that John Gath and the negroes under his care French Haggard and Nicholas Gentry Senr: be added to the road whereof John Estis is Surveyor

28 May 1745 O.S., Page 152

On the Motion of Abraham Venable Gent: Matthew Watson and Philemon Childers be added to his road

25 June 1745 O.S., Page 155

Hugh Hester is appointed Overseer of the road from Contrary to Christophers' Run including the bridge in the room of William Tate.

25 June 1745 O.S., Page 155

William Rice is appointed Surveyor of the highways in the room of Jeremiah Glen

25 June 1745 O.S., Page 155

Thomas Lipscomb on his petition is discharged from being any longer Surveyor of the highways from the County Line to Harris's branch and Godwin Trice is appointed in his room

25 June 1745 O.S., Page 155

Samuel Ragland on his Petition is discharged from being any longer Surveyor of the highways from Harris's branch to the north fork of Elk creek; and Thomas Kimbrow is appointed in his room

25 June 1745 O.S., Page 155

John Smithson on his Petition is discharged from being any longer Surveyor of the highways in this County and John Davis is appointed in his room.

25 June 1745 O.S., Page 155

Roger Thompson Gent. on his Motion is discharged from being any longer Surveyor of the highways from Owen's creek bridge to Deep creek; and David Shelton is appointed Surveyor in his room.

25 June 1745 O.S., Page 155

Ordered that the Several Surveyors of the highways in this County be continued as such (except those discharged today)

25 June 1745 O.S., Page 155

On the Motion of Thomas Prestwood for a road & c. The viewers this day made their report in these words "In obedience to the within order we have diligently viewed the road within mentioned & do believe the old way formerly granted to be the nearest & most convenient way to the church. Cha: Smith James Yancey. Archelus Yancey" Therefore it is Ordered that the old road be cleared and kept open for the use of the Publick & c.

25 June 1745 O.S., Page 155

Joseph Gooch Gent. and Thomas Jackson at the Petition of James Graves John Hill & others are appointed to view the way from the main road that leads to Charles Allens Ordinary to Henson's creek church, and report to the next Court where the most convenient road may be had

25 June 1745 O.S., Page 155

Roger Thompson Gent: Samuel Dalton and John Estis Surveyors of the highways in this County this day appeared to answer the Presentment of the grand Jury made against them for not keeping their roads in goods repair according to Law; and what they could Say in excuse for themselves being fully heard; it is Ordered that the Presentment be dismist.

* 27 August 1745 O. S., Page 164

On the Motion of William Waller Gent. on behalf of the Court of Orange to have a road cleared from the old Line of Orange County; at the end of a new road on the Top of the Ridge of the

Great Mountains through this County to the road near Martin Hackets leading to the falls of Rappahannock; it is Ordered that Robert Thomson Martin Hacket and John Keaton; do meet at Some convenient time before the next Court; and view the Same; and make report to the sd. next Court where the most convenient road may be had

26 November 1745 O.S., Page 173
William Hudson is appointed Surveyor of the highways in the room of Benjamin Johnson

* 26 November 1745 O.S., Page 173
To John MCcolly for Setting up Posts of directions to be paid by Joseph Bickley (sherif) 80 (lbs of tobo.)

* 28 January 1745 O.S., Page 175
Thomas Walker Gent. is appointed Surveyor of the highways from the top of the Little mountains in this County to the old Road

28 January 1745 O.S., Page 175
The hands belonging to Michael Holland & William Hudson; are added to William Hudsons Gang.

* 22 April 1746 O.S., Page 182
On the Petition of David Mills, John Fike, David Epperson, George Pearce, John Consolver, Thomas Ballard, Robert Thomson, Martin Hacket Henry Current; William Couhur William Land; Lewis Davis David Thomson John Bryson John Bryson Junr: Mosias Jones, Richard Bennet William Vauhan, Matthew Davis, Charles Mills, Ambrose Joshua Smith & Richard Searcy; for a road to be cleared on the ridge between the south & North Rivers to Todds pass; its Ordered that Benjamin Hensley John Davis & Samuel Dalton do lay off the said road & that the Petitioners & their Gangs clear the Same

22 April 1746 O.S., Page 182
Champness Terry upon his Motion is discharged from being any Longer Surveyor of the highways in this County and John Starke Gent: is appointed in his room

* 22 April 1746 O.S., Page 183
On the Petition of Benjamin Brown for a bridle way to be cleared from Doyles River down into Ennis's road; it is Ordered that Henry Bunch Joseph Keaton James Carney & Andrew Rea; or any three of them; meet at the said road and view the same; and report to the next Court; where the most convenient bridle way may be had

22 April 1746 O.S., Page 183

On the Petition of Richard Philips & others for a road to come out of Blalocks road into the ridge road; near the Widow Estis's; its Ordered that Thomas Ballard Smith & Philip Timberlake view the same; and report to the next Court were the most convenient road may be had

* 27 May 1746 O.S., Page 188

On the motion of William Waller Gent: in behalf of the Court of Orange to have a road cleared; from the Augusta County Line at the end of a new road on the Top of the ridge of the Great Mountains; through this County to the road near Martin Hackets; it is Ordered that Robert Thomson, Martin Hacket, Joseph Keaton and Henry Bunch, do meet & view the Same; and make report to the next Court

27 May 1746 O.S., Page 189

Ordered that the Overseer of the road from Harris's old Mill; to the road by the Church be summoned to appear at the next Court to answer the Presentment of the Grand jury this day made ag^t. him for not keeping his road in repair.

27 May 1746 O.S., Page 190

Ordered that the Overseer of the road; leading from the Main road by Thomas Sheltons to Henson's creek church be summoned to appear at the next Court to answer the Presentment of the Grand jury; for not seting up a Post of directions and for not keeping the said road in repair

* 27 May 1746 O. S., Page 190

Ordered that the Overseer of Turky Run road be Summoned to appear at the next Court to answer the Presentment of the Grand jury; for not keeping the s^d. road in repair

27 May 1746 O.S., Page 190

On the Petition of Gilbert Gibson to have a bridle way through Capt: Hollands Land; at Green Spring & Thomas Moremans Land, it is Ordered that John Smithson & Charles Moreman do view the same; and make report of their proceedings herein to the next Court

27 May 1746 O.S., Page 193

On the Motion of John Starke Gent. it is Ordered that Joseph Pulliam, Richard Holt Richard Wilborn & William Mcdowell & the Labouring male Tithables belonging to each of them; be added to the road whereof he is Surveyor

27 May 1746 O.S., Page 193

On the Motion of Samuel Thompson it is Ordered that Benjamin Bibb, Garfield Brown Archebald Carver Robert Anderson & Elisha Estis & the Labouring male Tithables belonging to each of them; be added to the road whereof he is Surveyor

24 June 1746 O.S., Page 194

On the Motion of Joseph Bickley Gent. it is Ordered that Henry Dickenson William Dickenson's Quarter; Barclay's quarter John Rhodes Dabney Pettus & the Labouring Male Tithables; belonging to Each of them; do clear the Church road; from the North River road to the Little River; and John Rhodes is appointed Overseer of the Same

24 June 1746 O.S., Page 194

Ordered that Benjamin Dumas Temperance Yancey Widow, Benjamin Harris Goodmans Quarter and William Crenshaw & the Labouring Male Tithables belonging to each of them; do clear the road from the Little River to the church & William Crenshaw is appointed Overseer of the Same

24 June 1746 O.S., Page 194

Ordered that Robert Garland Joseph Bickley, Archelus Yancey Daniel Maupine, James Yancey, Charles Smith, Henry Morris, Richard Brock John Morris John Tisdale William Chils & Charter Mitchel & the Labouring Male Tithables belonging to each of them do clear the road from Cleverius Duke's upper path to the fork of the road by Dabney Pettus's & James Yancey is appointed Overseer of the Same

* 24 June 1746 O.S., Page 194

Ordered that Samuel Dalton, Thomas Walker & John Meriwether do View from the Orange County Line to the Albemarle County Line below the Little Mountains & make report to the next Court; where the most convenient road may be had

24 June 1746 O.S., Page 195

William Hudson is appointed Overseer of the road from Gibsons Foard up to Samuel Dalton's path

* 24 June 1746 O. S., Page 195

Thomas Walker Gent: is appointed Overseer of the road from Samuel Dalton's path to the Top of the Turky Sagg Mountains; and it is Ordered that he have all the hands; that was formerly under Sam: Dalton & Francis Meriwether

24 June 1746 O.S., Page 195

Ordered that John Carr John Poindexter Joseph Terril, David Cosby Griffith Dickeson James Overton & John Saxon; and the Labouring male tithables belonging to each of them do clear the road from Elk creek road to the upper end of Christopher Smith's Orphans new plantation & Griffith, Dickeson is appointed Overseer of the Same

24 June 1746 O.S., Page 195

Ordered that Wm. Armistead Smiths Orphan's James Johnson Thomas Vowel William White, William Hendrick and John Kendrick and the Labouring male Tithables belonging to each of them; do clear the upper end of the road from Smiths Orphans New plantation; to the round slash between the branches of contrary & North East creek & James Johnson is appointed Overseer of the Same

24 June 1746 O.S., Page 195

Ordered that Robert Estis Moses Estis, Benjamin Wright; John Snelson, James Powers, Alexander Freeman, Edward Stringer John Say Henry Milles, James Roach & David Roach; and the Labouring male Tithables belonging to each of them; do clear the road from the round slash into the Courthouse road & Robert Estis is appointed Overseer of the Same

24 June 1746 O.S., Page 195

Thomas Ballard Smith & c. being appointed to view the way out of Blalocks road into the ridge road near the widow Estis's; and to report where the most convenient road might be had; and now reporting that they find; a road may be conveniently carried to the place aforesaid; it is therefore Ord^d: that a road be laid off and clear'd accordingly; and Thomas Ball^d: Smith is appointed Overseer thereof and the Labouring male Tithables belonging to Beverly Randolph Gent: John Winston John Powers Sylvanus Jouett & Richard Philips are allotted the said Overseer to assist in clearing the s^d: road & keeping the Same in good repair

* 24 June 1746 O.S., Page 196

Ordered that the following persons be added to the road from the Turkey Sagg to the road above the Mountains, Viz: Joseph Wood Richard Briggs John Dowel Thomas Collins Ambrose Smith & William Hall & the Labouring male tithables belonging to each of them & John Maccaulay is appointed Overseer of the Same

* 24 June 1746 O.S., Page 196

Richard Hammack is appointed Overseer in the room of John Maccaulay

22 July 1746 O.S., Page 197

John Tate is appointed Surveyor of the highways from Hensons Creek church to the County Line in the room of Thomas Hardy

* 22 July 1746 O.S., Page 197

Ordered that Richard Durrett; with the assistance of his own Gang, Richard Hammacks, John M^{c}caulays Thomas Ballards & William Courseys make a bridge over Pritty's Creek in the Most convenient place between Capt: Smith's plantation & his mill.

* 26 August 1746 O.S., Page 198

On the Petition of John Brison for a bridle way from the County Line to Martin Hackets; it is Ordered that Robert Thomson Lewis Davis & David Milles; do view the same & report to the next Court; were the most convenient road may be had; doing the least prejudice to any person

* 26 August 1746 O.S., Page 198

Ordered that John Ennis be added to the other persons appointed to view a road on the petition of Henry Downs

26 August 1746 O.S., Page 200

Charles Moreman & John Smithson being appointed to view the way through Capt. Hollands & Thomas Moreman land to Gilbert Gibsons mill & now reporting that the way aforesaid will not be prejudicial to anyone therefore it is ordered that a bridle way be laid off and cleared accordingly

23 Sept. 1746 O.S., Page 203

Grand=Jury	Presentment
ag^t:	Dismist & Richard Yancey is
The Overseer of the road from	appointed Overseer thereof
Harris's old Mill to the Church	

* 23 Sept. 1746 O.S., Page 208

The Petition of John Brison for a bridle way is continued till the next Court

23 Sept. 1746 O.S., Page 208

The Petition of Henry Downs for a bridle way is continued till the next Court

25 Nov. 1746 O.S., Page 211

Philip Timberlake is appointed Overseer of Indian Creek road in the room of Richard Palmer

25 Nov. 1746 O.S., Page 212

To Richmond Terril for Setting up a Post of directions 80 (lbs. tobo.)

25 Nov. 1746 O.S., Page 212

To William Allen for setting up a post of directions 30 (lbs. tobo.)

* 27 J an. 1746 O. S., Page 217

The Petition of John Bryson for a bridle way is continued till the next Court

27 Jan. 1746 O.S., Page 218
   The Petition of Henry Downs for: a road continued till the next Court

* 24 Feb. 1746 O.S., Page 219
   Ordered tha Maj^r: Meriwether Micajah Clark, Christopher Clark, Bolen Clark & William Harris; and the Labouring male Tithables belonging to each of them do clear the road from Albemarle Line to Colo: Lewis's mill; and Micajah Clark is appointed Overseer of the Same

* 24 Feb. 1746 O.S., Page 219
   Ordered that Col^o: Lewis's people & Anthony Pates do clear the Road from Colo: Lewis's mill to the new road; and Robert Lewis Gentleman is appointed Overseer of the same

* 24 Feb. 1746 O.S., Page 219
   Ordered that James Meriwether, Col^o: Lewis Qu^r: Jno: Moore, George Eastham and five of Samuel Daltons hands; do clear the road from the new road to the Orange County Line & John Meriwether is appointed Overseer of the same.

* 24 Feb. 1746 O.S., Page 220
   On the Petition of John Brison for a bridle way through Josias Woods' land; into Martin Hackets road; its granted

* 24 Feb. 1746 O.S., Page 220
   Ordered that the road whereof Richard Durret is Overseer be divided into two Precincts; Viz^t. from the bridge above Capt. Smith's plantation, to the fork of the road; and Thomas Henry is appointed Overseer of the Same. And the other from the fork of the road; to the County Line; the said Durret is appointed Overseer of the Same

28 April 1747 O.S., Page 224
   Ordered that Abraham Allen be Overseer of the road; called Milles's road in the room of Thomas Ballard.

28 April 1747 O.S., Page 224
   Ordered that the hands belonging to John Powers be discharged from Serving on the road whereof Thomas Ballard Smith is Surveyor

* 28 April 1747 O.S., Page 224
   Ordered that John Thomson be added; to the order for clearing a road from the Augusta County Line

26 May 1747 O.S., Page 227
   Ordered that the Surveyor of the road from Christophers run to Mr. Starkes road be Summoned to appear at the next Court to answer the Presentment of the Grand jury made ag^t: him for not Setting his road in repair

26 May 1747 O.S., Page 227
      Ordered that Benjamin Henson be Summoned to appear at the next Court to answer the Presentment of the Grandjury made ag^t: him for not Setting up a Post of directions on his road leading to Elk Creek

26 May 1747 O.S., Page 227
      Ordered that Thomas Ballard Smith be Summoned to appear at the next Court to answer the Presentment of the Grandjury made ag^t: him for not Setting up a Post of directions according to Law

* 26 May 1747 O. S., Page 227
      Ordered that the Surveyor of Richard Hammacks road; be summoned to appear at the next Court to answer Presentment of the Grandjury made against him for not Setting up a Post of directions according to Law

26 May 1747 O.S., Page 227
      Ordered that the Surveyor of the main County road be summoned to appear at the next Court to answer the Presentment of the Grandjury made against him for not keeping the Lower Bridge over Bever Creek in repair.

23 June 1747 O.S., Page 232
      Ordered that the Road whereof John Gentry was formerly Surveyor be divided thus from the main road by the church down to the path, where Mr. Barrets path crosses to John Cosby's Quarter; and James Nuckolls is appointed Surveyor of the Same; and the Labouring male tithables of John Cosby's upper quarter Charles Barrets, Benjamin Henson & James Fears are allotted to the said Surveyor to keep his road in repair; and Overton Harris is appointed Surveyor J from the path that crosses to Mr. Cosbys; and the Labouring male tithables of John Cosby's Lower quarter, Benjamin Bibb, Richard Yancey & William Brown are allotted to the s^d: Surveyor to keep I his road in repair

23 June 1747 O.S., Page 232
      Robert Estis is appointed Overseer of the road from Harris's creek to Elk creek in the room of Thomas Kimbrow.

23 June 1747 O.S., Page 232
      David Cosby, William Saxon & Griffith Dickenson, on the Petition of John Poindexter Gent: are appointed to view the way from Elk Creek up to the Courthouse into the road that leads from Harris's old Mill; to the new church and report to the next Court where the most convenient road may be had

23 June 1747 O.S., Page 232
      Charles Bickley on his Motion is discharged from being any Longer Surveyor of the road from Elk creek up to contrary including the bridge and James Winston is appointed in his room

23 June 1747 O.S., Page 232
      Daniel Williams is appointed Surveyor of the highways from Indian creek to the old road including the bridge in the room of Abraham Venable Gent.

23 June 1747 O.S., Page 232

Thomas Paulet is appointed Surveyor of the highways from Indian Creek up to the head of the road in the room of Abraham Venable Gent

23 June 1747 O.S., Page 232

Ordered that all Surveyors of roads in this County (except those discharged today) be continued as Such for this Year

* 23 June 1747 O.S., Page 234

The Petition of Benjamin Brown p^r. a bridle way is Continued till the next Court

23 June 1747 O.S., Page 234

Henry Downs Gent. p^r. a road Continued till the next Court

23 June 1747 O.S., Page 234

Ordered that James Watson & his hands be added to the road whereof Benjamin Henson is Surveyor.

28 July 1747 O.S., Page 235

Ordered that all the hands belonging to M^r. Venables road above north East creek be allotted to the road whereof Thomas Paulet is Surveyor.

28 July 1747 O.S., Page 235

Ordered that all the hands belonging to M^r. Venables road below north, East creek be allotted to the road whereof Daniel Williams is Surveyor.

28 July 1747 O.S., Page 235

Ordered that Roger M^{c}guire be Overseer of the road from Cub Creek bridge to the County line.

28 July 1747 O.S., Page 236

Ordered that Thomas Gresham be Overseer of the road that turns out of the road just above Smith's Quarter, thence to keep the ridge between little river and Contrary into the road from Harris's mill a little below the Church & that the Labouring male Tithables belonging to Snelsons Quarter, William White, William Hendrick James Overton, William Saxon, Griffith Dickenson, David Cosby, Joseph Terrils quarter, John Poindexter, Armisteads quarter & James Winston be allotted to assist in keeping the said road in repair.

28 July 1747 O.S., Page 236

  The Court having under their consideration, the building of a Bridge over fork creek in this county; are unanimously of opinion that there ought to be a bridge over the Same; Therefore it is Ordered that Joseph Fox, Joseph Shelton, & Roger Thompson Gent: do agree with workmen for the building of the Same

* 25 August 1747 O.S. ,Page 237

  Thomas Walker & Joseph Martin Gent: on the Petition of divers inhabitants of this County; are appointed to view the way; were Mr. Martins roling path comes into the road over the north river; below W^m. Carrs quarter into Buck mountain road; and report to the next Court

25 August 1747 O.S., Page 239

  The Petition of Henry Downs for a road Continued till the next Court

* 25 August 1747 O.S., Page 239

  The Petition of Benjamin Brown for a road Continued till the next Court.

27 Oct. 1747 O.S., Page 244

  The Petition of Henry Downs for a road is continued till the next Court

24 Nov. 1747 O.S., Page 248

  On the Motion of Robert Anderson for a road from hiccory creek bridge to the Lower Church; its Ordered that Samuel Thompson, Robert Anderson Robert Estis & George Thomason or any three of them; do meet and view where the most convenient road may be had & report their proceedings herein to the next Court.

* 24 Nov. 1747 O.S., Page 248

  Ordered that John M^c colly turn the road that leads from Albemarle County to Orange County

* 24 Nov. 1747 O.S., Page 248

  On the Petition of Divers Freeholders of this County for a road from Mr. Martins roling path over the north River below William Carrs' quarter into Buck mountain road; its Granted, and Ordered that Thomas Henry's Gang, Jno. Mccollys, Rich^d. Hammack & William Coursey's do meet & open the Same & David Watts is appointed Overseer thereof.

24 Nov. 1747 O.S., Page 248

On the Petition of Divers inhabitants of this County to have a road to pass out of Elk creek near Mr. Smith's quarter; running thence into the road near Benjª: Bibbs; its Ordered that John Rhodes, Griffith Dickenson, William Sexton, & James Overton or any three of them meet & view the Same & make report to the next Court.

25 Nov. 1747 O.S., Page 248

	To Thomas Balld: Smith for Setting up a post of directions		20
	" John Storke	"	20
	" Thomas Gresham for two posts of directions		40
	" John Estis	"	40
*	" Thomas Henry	"	20
*	" Richd. Hammack	"	40

23 Feb. 1747 O.S., Page 249

Robert Davis is appointed Surveyor of a road in this County in the room of John Davis

23 Feb. 1747 O.S., Page 249

Ordered that William Hudson be Overseer of the road from Gibsons foard to Daltons path & that the hands of M^{rs}. Morris & James Flanacan be added to his Gang

23 Feb. 1747 O.S., Page 250

On the Motion of Samuel Thompson it is ordered that he turn the road according to M^r. Starks directions

23 Feb. 1747 O.S., Page 250

On the Petition of divers freeholders of this County for a Bridge to be built over the Northanna River, its continued till the next Court.

23 Feb. 1747 O.S., Page 250

On the Motion of James Winston; it is ordered that his hands be taken off of the road whereof Thomas Gresham is Surveyor

23 Feb. 1747 O.S., Page 250

Ordered that Richmond Terril & Richard Thurman & their hands be taken of, off the Church road & added to the road whereof John Tate is Surveyor.

23 Feb. 1747 O.S., Page 250

Adam Jones is appointed Surveyor of a road in this County in the room of Richmond Terril.

23 Feb. 1747 O.S., Page 250

On the Petition of Richard Walker & others for a bridle path; from Walkers ordinary to Thomas Ballard Smiths Mill; it is Ordered that Thomas Jackson & Jeremiah Glen do view the Same; and make report to the next Court

23 Feb. 1747 O.S., Page 250

On the Petition of Divers inhabitants of this County for a bridle path; to turn out of the road between Christophers run & Tommahuck; it is Ordered that William White & Robert Estis do view the same; and make report to the next Court.

23 Feb. 1747 O.S., Page 250

On the Motion of Robert Anderson it is Ordered that a road be laid off and cleared from hiccory creek bridge to the Lower Church & that it be divided at James Roaches plantation & Robert Anderson is appointed Overseer of the upper part, thereof & all the Labouring male Tithables belonging to Joseph Pulliams quarter, Samuel Thompson, John Starke, John Poindexter's quarter, Robert Estis Junr: William M^{c}doel, Richard Wilburn, Temples quarter, Willm. White's quarter & Elisha Estis are allotted to the s^d. Anderson to keep the Same in repair. And Alexander Freeman is appointed Overseer of the Lower part thereof & all the Labouring male Tithables belonging to Garfield Brown, James Roach, David Roach James Powers Qu^r: Snelsons quarter, Henry Milles, Edward Stringer, John Garland & William Bennet are allotted to the s^d. Freeman to keep his road in repair.

26 April 1748 O.S., Page 265

The Petition of Henry Downs for a road & c. not being returned; is Dismist

26 April 1747 O.S., Page 265

On the Petition of Sundry Freeholders & inhabitants of this county to have a bridge Erected over the Northanna or Pamunkey river, which devides this & the County of Spotsylvania; at Some convenient place between the mouth of great rocky creek & Franks' run, the petition is granted. And it is Ordered that Charles Barret, Robert Harris, John Carr, & Joseph Bickley Gent: or any two of them, do meet Commissioners to be appointed by the said Court of Spotsylvania & with them do agree and fix on the most convenient place within the Limits afores^d: for erecting the s^d. bridge & make report of their proceedings to the next Court.l.

* 24 May 1748 O. S., Page 266

John Goodall is appointed Overseer of the road from Buck mountain creek to the upper fork of rocky creek

24 May 1748 O.S., Page 267

William Thompson is appointed Overseer of the road from the North fork of Willis's road to Highgate./.

24 May 1748 O.S., Page 267

The view of the bridge over the Northanna River being returned it is Ordered that Charles Barret, Robert Harris, John Carr &

Joseph Bickley Gent. or any two of them do agree with Workman for the building of the Same between this & the next Court./.

28 June 1748 O.S., Page 272
Thomas Gresham is discharged from being any Longer Surveyor of a road & William Hendrick is appointed in his room./.

28 June 1748 O.S., Page 279
The Persons appointed by this Court to meet Commissioners appointed by the Court of Spotsylvania returned that they have with the said Commissioners agreed with Benjamin Davis of the County of Spotsylvania to Erect a bridge at the place mentioned in the former return for the Sum of thirty Seven pounds ten Shillings which is to be paid to him by both the s^d. Counties; in proportion to the number of Tithes in each of them, the said Davis is to maintain the said bridge for Seven years and the said Comconrs of Spotsylvania is to take bond for the performance of the Same; which is ordered to be recorded.

END OF ORDER BOOK 1742-1748

INDEX - LOUISA COUNTY ROAD ORDERS

Note: This index is arranged by subject: Personal Names; Bridges; Churches; County Lines; Courts (of other Counties); Fords; Landmarks, Buildings, Geographical Formations; Mills; Plantations; Rivers, Creeks, Branches, Swamps, etc.; Roads; Road Surveyors Gangs; Quarters; Signposts

Personal Names
Abraham Allen, 23
Charles Allen, 9
William Allen, 6, 12[2], 13[4], 14, 22, 23
Robert Anderson, 20, 26[2], 28[3]
Mr. Anderson, 9
Lancelot Armstrong, 6
Thomas Ballard, 18, 21, 23
Charles Barret, 5, 6, 13, 14[2], 24, 28[2]
Mr. Barrets, 24
Richard Bennet, 8, 18
William Bennet, 28
Benjamin Bibb, 20, 24, 27
Henry Bibb, 12, 14
Charles Bickley,10, 12, 14, 24
Joseph Bickley, 5, 13[2],18, 20[2], 28
William Biggars, 7
Richard Briggs, 21
Richard Brock, 20
John Bryson, 18, 22[3], 23
John Bryson Jun[r]:, 18
Benjamin Brown, 7, 18, 25, 26
Garfield Brown, 20, 28
William Brown, 8, 24
Richard Bullock, 7, 12[2]
Henry Bunch, 5[3], 18, 19
James Carney, 5, 18
Gideon Carr, 6, 9,16
John Carr, 8, 9, 10, 11, 14, 15, 20, 28[2]
Archebald Carver, 20
Philemon Childers, 16
John Chiles, 7
William Chils, 20
Thomas Christmas, 6, 12
Bolen Clark, 23
Christopher Clark, 15, 23
Joseph Clarke, 8
Micajah Clark, 14, 23[2]
Cap[t]: Clark, 14
Thomas Collins, 21
John Consolver, 18
John Cook, 5, 10, 11[2]
David Cosby, 5[2], 7, 14[3], 20, 24, 25
John Cosby's, 24
Mr. Cosbys, 24
William Couhur, 18

William Coursey, 6,15,16, 21, 26
William Crenshaw, 20
Henry Current, 18
Peter Curry, 8
Samuel Dalton, 5, 16, 17, 18, 20[2], 23
William Davenport, 13[3]
Benjamin Davis, 29
George Davis, 16
John Davis, 10, 11[2], 17, 18, 27
Lewis Davis, 18, 22
Matthew Davis, 18
Richard Davis, 15, 16
Robert Davis, 27
William Davis, 5
Griffith Dickenson, 8, 11, 14, 20[2], 24, 25, 27
Nathaniel Dickenson, 9
Henry Dickerson, 20
John Dickerson, 9
Doct[r]. Dixon, 6
John Dowell (see also McDowell), 16, 21
Henry Downs, 22[2], 23, 25, 26[2], 28
Cleverius Dukes, 9, 20
Benjamin Dumas, 5, 20
Richard Durret, 10, 15, 21, 23[2]
George Eastham, 23
Robert Ellis, 14
John England, 11, 13[3]
Stephen English, 16
John Enes, 5
John Ennis, 22
David Epperson, 8, 18
Elisha Estis, 20, 28
James Estis, 9
John Estis, 6, 7, 8, 10, 16[2], 17, 27
Moses Estis, 14,21
Richard Estis, 8
Robert Estis, 8, 14, 21[2], 24, 26, 28[2]
Widow Estis's, 19, 21
James Fears, 24
John Fike, 18
Joseph Fox, 13[2], 14, 26
James Flanacan, 27
James Frances, 5
Alexander Freeman, 21, 28[2]
John Garland, 28
Robert Garland, 7, 20
John Garrison, 14
John Gath, 16
John Gentry, 7, 24
Nicholas Gentry, 6, 7, 15
Nicholas Gentry Sen[r]:, 16
Samuel Gentry, 15
Gilbert Gibson, 19
Jeremiah Glen, 11, 16, 27

Mr. Glen, 10, 11
Forest Green, 7
Thomas Gresham, 14, 25, 27[2], 29
Joseph Gooch, 17
William Gooch, 6
John Goodall, 28
James Graves, 17
Wm. Gray, 6
John Hacket, 5
Martin Hackett, 5, 8[2], 18, 19, 22
French Haggard, 7, 16
Richard Haggards', 16
William Hall, 21
John Hammack, 16
Richard Hammack, 16[2], 21, 24, 26, 27
Thomas Hardy, 8, 21
John Harlow, 14
Benjamin Harris, 20
Jn°: Harris, 9
Overton Harris, 24
Robert Harris, 13, 28[2]
William Harris, 6, 23
John Hendrick, 12
William Hendrick, 14, 21, 25, 29
Thomas Henry, 23, 27
Benjamin Hensley, 9[2], 16, 18
John Hensley, 16
Samuel Hensley, 16
Benjamin Henson, 24[2], 25
John Henson, 6
Hugh Hester, 16
John Hill, 17
James Hoderurds, 16
Michael Holland, 18
Capt: Hollands, 22
Richard Holt, 19
William Hudson, 15, 18[2], 20, 27
Ellis Hughes's, 7
Thomas Jackson, 11, 17, 27
Mr. Jackson, 10, 11
Anne Johnson, 8[2]
Benjamin Johnson, 6, 9, 18
James Johnson, 14, 21[2]
Adam Jones, 9[2], 27
Mosias Jones, 10, 11, 18
Thomas Jones, 9
Sylvanus Jouett, 21
James Keaton, 5
John Keaton, 18
Joseph Keaton, 5, 18, 19
John Kendrick, 6, 21
Thomas Kimbrow, 17, 24
William Land, 18
Edward Lane, 7

Robert Lewis, 14, 15, 23
Thomas Lipscomb, 7, 16
John McCauley (McCcolly,etc.), 9, 15, 16[2], 18, 21[3], 26 [2]
John M^{c}dowell (see also Dowell), 14, 28
William M^{c}dowell, 19
Samuel M^{c}gehee, 12[2], 13[3]
Roger M^{c}guire, 25
Alexander MacKoy, 5
Richard Madors, 8
Joseph Martin, 9[2], 16, 26
Mr. Martins, 26
Daniel Maupine, 20
Beasell Maxwell 5
William Maybe (Maib), 6, 16
James Merideth, 10, 11, 14
David Meriwether, 10, 11
Francis Meriwether, 20
James Meriwether, 23
John Meriwether, 10, 12, 15[3], 20, 23
Nicholas Meriwether, 15
Thomas Meriwether, 14, 15
Madame Meriwethers, 15
Maj^r: Meriwether, 23
Mr. Meriwethers', 14
Charles Mills, 18
David Mills, 10, 11[2], 18, 22
Henry Mills, 8, 9, 21, 28
Charter Mitchel, 20
John Moony, 16
Timothy Moony, 16
Jno: Moore, 23
Charles Mooreman, 5, 7, 19, 22[2]
Thomas Moreman, 22
Henry Morris, 20
John Morris, 20
Mrs. Morris, 27
John Moss, 9
James Nuckols, 5, 6, 7, 24
William Oens, 5
William Ogilvie, 6, 10
James Overton, 14, 20, 25, 27
Richard Palmer, 10, 22
Anthony Pates, 23
Thomas Paulett, 9, 25[2]
George Pearse, 8, 18
Dabney Pettus, 20[2]
Stephen Pettus's, 5
Richard Philips, 19, 21
John Poindexter, 10, 11[2], 14, 20, 24, 25
James Powers, 21
John Powers, 21, 23
Thomas Prestwood, 16, 17
Joseph Pulliam, 19

Mr. Thompson, 10
Richard Thormond, 9[2], 27
Philip Timberlake, 19, 22
John Tisdale, 20
Godwin Trice, 16
William Vauhan, 18
Abraham Venable, 7, 16, 24, 25
Thomas Vernon, 5
Thomas Vowel, 21
Anthony Waddy, 6
Richard Walker, 13, 27
Thomas Walker, 12, 15[3], 18, 20[2], 26
Doct[r]. Walker, 10
William Waller, 17, 19
James Watson, 8, 25
Matthew Watson, 16
David Watts, 26
George Webb, 10, 11
Mr. Webb, 11[2]
Samuel Wells, 5, 14[2]
William White, 14, 21, 25, 28
James Whitlock, 6
Richard Wilburn, 8, 19, 28
Daniel Williams, 24, 25
John Wilmer, 16
James Winston, 24, 25, 27
John Winston, 21
John Wood, 16
Joseph Wood, 21
Josias Woods', 23
David Woodroof, 6
Benjamin Wright, 21
Richard Wright, 7
Archaelus Yancey, 16, 17, 20
James Yancey, 16, 17, 20 [2]
Richard Yancey, 6, 7, 22, 24
Temperance Yancey, 20

Bridges:
Bever Creek, 24
Camp-creek, 5
Cattale, 11
Christophers Run, 6, 16
contrary creek, 6, 12, 24
Cub Creek, 25
fork creek, 26
hiccory creek, 26, 28
Hudson's bridge, 5[2]
Indian creek, 24
North Anna river, 7, 27, 28[2]
Owens Creek, 7, 8, 17
Pritty's Creek, 16, 21, 23

Churches:
Church, 11, 12, 17, 19, 22, 25
Henson's Creek Church, 10, 13[2], 17, 21, 19
Lower Church, 26, 28
new church, 16, 24
old chaple, 14

County Lines
County Line (Louisa), 6, 8[2], 9, 12, 16, 21, 22, 23, 25
Albemarle County Line, 20, 23
Augusta County Line, 19, 23
Goochland County line, 6, 14
old Line of Orange County, 17
Orange County Line, 6, 8, 20, 23

Courts (of other Counties):
Hanover, 5[2], 6
Orange, 17, 19
Spotsylvania, 28[2], 29
Commissioners of, 28, 29

Fords:
Foard, 7, 12
Gibsons foard, 20, 27
Goodalls Lower ford, 10
Rocky Foard, 7, 12

Landmarks, Buildings, Geographical Formations:
Charles Allens Ordinary, 17
burnt School house, 13
Chesnut Ridge, 10, 12, 15
Courthouse, 24
Great Mountains, 19
green Mountain, 14
Little mountains, 18, 20
Mountains, 5, 21
John Estes's Race Ground, 5[3], 10
ridge between little river and Contrary, 25
ridge between south & North Rivers, 18
Todds pass, 18
Top of the mountains, 15
Turkey Run Saggs, 8
Turkey Sagg, 21
Turkey Sags, 15
Turky Sagg Mountains, 20
Capt: Thomas Meriwethers Smiths Shop, 14
Walkers ordinary, 27
Public Warehouse, 15

Mills:
Gilbert Gibsons mill, 22
Harris's old Mill, 19, 22, 24
Col°: Lewis's mill, 23
Ambrose Joshua Smiths Mill, 16

Thomas Ballard Smiths Mill, 27
Capt. Smith's mill, 21
Old Dam, 5

Plantations:
Bickleys, 13
Richard Hammack's, 16
Stephen Pettus's, 5
James Roaches, 28
Christopher Smith's Orphans new, 20, 21
Capt: Smith's, 21, 23
Smith's, 6[2]

Rivers, Creeks, Branches, Swamps, etc.:
Bever Creek, 11
Bever damm fork, 14
Buck mountain Creek, 5[2], 8, 15, 28
Cabbin branch, 13
Christophers Run, 6, 7, 9, 11, 16, 23, 28
Contrary creek, 6[2], 9, 11, 12, 14, 16, 21, 25
Cub Creek, 25
Deep creek, 17
dirty Swamp, 15, 16
Doyles River, 18
Ducking hole, 11
Elk Creek, 6, 8, 11, 12[2], 14, 17, 24[3], 26
East (creek), 11
falls of Rappahannock, 18
fork creek, 16, 26
Foster's Creek, 11
Fox branch, 7
Franks' run, 28
Gilbert Gibsons Mill creek, 22
Gold Mine, 11
Green Spring, 19
Harris's branch, 16, 17
Harris's creek,   24
Horse-pen swamp, 5
Hickory Creek, 7
Hinsons Creek, 13
Indian creek, 24, 25
Linches river, 8
Little creek, 10
Little River, 7, 8, 9, 12, 20, 25
Long Creek, 12
North River, 9, 18, 26[2]
North Anna River, 7, 27, 28
North East creek, 21, 25[2]
Owen's Creek, 8, 16
Pamunkey river, 28
Pritty's Creek, 16, 21
Rocky branch, 7, 12
Great Rocky Creek, 7, 12, 28
Rocky Creek, 5, 28

Roundabout Creek, 11
round slash, 21(2)
south River, 18
Tommahuck (creek), 28
Turkey-run, 6
South Side of the River, 8
North side of the North fork of James River, 6
River, 5

Roads:

Albemarle Courts to Orange County, 26

from Albemarle Line to Col^o: Lewis's mill, 23

from Augusta to County Line on Top of Great Mountains, 19, 23

from M^r.Barret's Church path, 7, 12

from M^r.Barrets path, 24

road near Benj^a: Bibbs, 27

Blalocks road, 11, 19, 21

road out of Blalocks road into ridge road, 19, 21

Buck mountain road, 26$^{(2)}$

Henry Bunches road, 15

Christmas's road, 8, 13

road from Christophers run to Mr. Starkes road, 23

bridle path to turn out of the road between Christophers run and Tommahuck, 28

Church Road, 8, 19, 20, 27

Capt: Clark's road, 14

bridle way from County Line to Martin Hackets, 22

Courseys Road, 5,15

Courthouse road, 21

path that crosses to Mr. Cosbys, 24

road from Cub Creek bridge to the County Line, 25

Samuel Dalton's path, 20$^{(2)}$, 27

road from Samuel Dalton's path to top of Turky Sagg Mountains, 20 road that crosses at Dickerson's Mill, 16

bridle way from Doyles River down into Ennis's road, 18

Cleverius Duke's upper path, 20

road from Cleverius Duke's upper path to the fork of the road by Dabney Pettus's,   20

road from Elk creek road to the upper end of Christopher Smith's Orphans new plantation, 20

road to pass out of Elk creek near M^r.Smith's quarter, 26

Elk creek Road, 12, 20

road leading to Elk Creek, 24

road from Elk Creek to the Church, 12

road from Elk creek to contrary, 14, 24

Ennis's road, 18

Road from Fork Creek to Deep Creek, 8

Gibson's road,   15

road from Gibsons Foard to Dalton's path, 20, 27

Martin Hackets road, 23

road near Martin Hackets, 19

road near Martin Hackets leading to the falls of Rappahannock, 18

Richard Hammacks road, 24

road from Richard Hammack's Plantation, 16

William Harris's Cartway, 5

road from Harris's creek to Elk creek, 24

road from Harris's mill, 25

road from Harris's old Mill, 19

road from Harris's old Mill to the Church, 22

road that leads from Harris's old Mill to the new church, 24

Bridle road to the church at Henson's creek, 13

(road) from Hensons Creek church to the County Line 21

road above the Mountains, 21

North River Road, 10, 11, 14, 20

Road out of the north River road, 14

road over the north river, 26

road by Dabney Pettus's, 20

fork of the road by Dabney Pettus's, 20

fork of the road (near Prittys Creek), 23[2]

Ridge Road, 9, 19, 21

road from the round slash into the Courthouse road, 21

road from Roundabout Creek to Fosters Creek, 10

road from Smiths Orphans New plantation to the round slash between the branches of contrary & North East creek, 21

road just above Smith's Quarter, 25

road that turns out of the road just above Smith's Quarter, 25

Road from the mouth of Starke's branch to Aylett's branch, 7

Mr. Starkes road, 23

Champnes Terry's road, 7

Three Notch road, 9

Turky Run road, 19

road through Turkey Sags, 15

road from the Turkey Sagg to the road above the Mountains, 21

M[r].Venables road, 25[2]

bridle path from Walkers ordinary to Thomas Ballard Smiths Mill, 27

bridle way through Josias Woods land; into Martin Hackets road, 23

road from the North fork of Willis's road to Highgate, 28

Willis's road, 28

Arche[s]: Yancey Path, 13

Bridle road, 13[2]

new road, 23[2]

road from the new road to the Orange County Line, 23

new road on top of Great Mountain, 19

new road on the Top of the Ridge of the Great Mountains, 17

old Road, 18, 24

road on the ridge between south & North Rivers to Todds pass, 18

roling path, 26

Road Surveyors' Gangs:
Allen's Gang, 13
Thomas Ballard's Gang, 16
William Coursey's Company, 6
Thomas Hardy's Gang, 9
Thomas Henry's Gang, 26
William Hudsons Gang, 18
Mr. Thompsons Gang, 8
Mr. Thompson's precinct, 7

Quarters:
Armisteads, 14, 25
Barclay's, 20
William Carr's, 26(2)
John Cosby's, 6, 24
John Cosby's upper quarter, 24
John Cosby's Lower quarter, 24
William Dickenson's, 20
Doct^r. Dixon's, 6
Fork Quart^r., 6
Goodmans, 20
Governor's, 14
Hower's, 14
Col^o: Lewis, 23
Maj^r: Morris's, 7, 14
James Overton's, 11
Overton's, 10
John Poindexter's, 14, 28
James Powers, 28
Joseph Pulliams, 28
John Shelton's, 14
Smith's, 14, 25, 26
Snelsons, 25, 28
Temples, 28
Joseph Terrils, 25
Willm. White's, 28
M^r. Winstons Two Quarters, 14

Quarters where Christopher Berryman is Overseer, 6

Signposts, 12, 18, 19, 22$^{(2)}$, 24$^{(3)}$, 27$^{(5)}$

www.ingramcontent.com/pod-product-compliance
Lightning Source LLC
Chambersburg PA
CBHW080537090426
42733CB00015B/2606